Melanin: Energy from Light and the Souls Ascent

Melanin Research Foundation

Published by The Book Shed

978-1-943392-01-8

Contact: coopertonnn@gmail.com

DEDICATION

To all seekers of the New World

Figure 1 Extracted Melanin recording a .737 Voltage in Sunlight

CONTENTS

The mitochondrion is the metabolic hub of the cell. Protons get sequestered to the inner membrane, in order to accumulate an electric gradient which activates ATP synthase to form ATP. This process is known to be ultimately fueled by the combustion of biomolecules such as glucose. NADH and $FADH_2$ harbor a majority of this energy, and deliver protons (H^+) and electrons to the mitochondrion to establish the electrical gradient. This process has been known for a while and is how energy, in the form of ATP, is made available. This process of aerobic respiration and glycolysis is about 50% efficient. Taking into account the scarcity of food, this process is far from perfect. It would be naïve to think that the human did not evolve to metabolize energy from a more abundant resource – the sun.

Photopigments absorb radiation. There is a vast array of photopigments, most notable is chlorophyll which is involved in plant photosynthesis. There are other organisms that also generate energy from the sun. Aphids demonstrated photometabolic capabilities [1]. The pectin of birds, packed with the photopigment melanin, was shown to directly convert light for metabolic use. Melanized fungus have been shown to use radiation for growth [2]. Since humans have melanin also, this would be the most likely candidate for human photometabolism. Melanin is like chlorophyll, except much more effective at absorbing sunlight. That is why melanin is black, because it absorbs the entire spectrum of visible light and most invisible wavelengths of electromagnetic radiation.

In humans, melanin is a pigment present in the skin (eumelanin), the brain (neuromelanin), and hair (pheomelanin). Dark skin indicates its presence, and a sun tan is the result of melanin accumulation. Melanin is "...capable of dissipating >99.9% of absorbed UV and visible radiation..." from the sun [3]. In hindsight, it seems intuitive that the human would have evolved to harness energy from the sun. Recently, it has been discovered by Arturo Solis Herrera and his crew that melanin has the capability to split water into diatomic hydrogen and oxygen with sunlight as a catalyst [4]. These implications are revolutionary. I hope to see Dr. Herrera receive a Nobel Prize in the near future. This process of splitting water, hydrolysis, absorbs about 286kJ of energy per mol of reacted water.

$$H_2O_{(l)} \longleftrightarrow H_{2(g)} + 1/2\ O_{2(g)}$$

$$\Delta G = -237.1\ \text{kJ/mol}$$

$$\Delta H = -286\ \text{kJ/mol}$$

Figure 2. The splitting of the water molecule and the respective energy absorption values

The hydrogen gas is the locale of this energy absorption. After it reduces NAD^+ or FAD, H_2 becomes immediately available to be used in the electron transport chain and help accumulate the proton gradient within the mitochondrion's inner membrane. theoretically much more efficient than glycolysis and pyruvate oxidation for NADH and $FADH_2$ production. Energy from the sun is absorbed and used to split the water molecule into a metabolically useful form. This makes sense of melanin granules adhering to mitochondria as seen to the right.

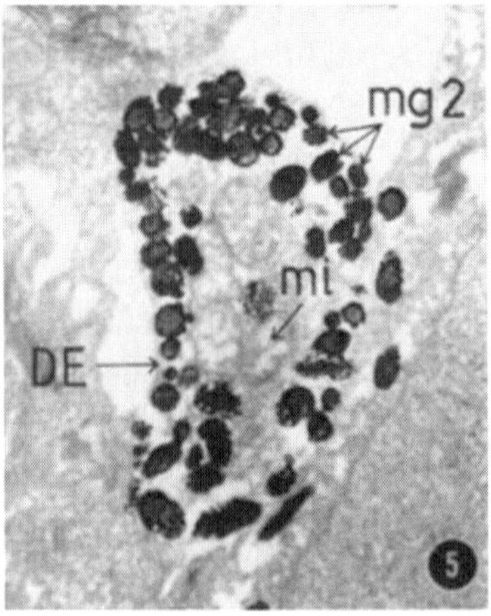

During the early stages of melanogenesis, the mitochondrion physically adheres to the developing melanin-packed vacuole. This is also why mitochondria are involved with melanin production [5]. Melanin granules were shown to contain succinate dehydrogenase, otherwise known as complex II of the electron transport chain. Complex II is the same enzyme present in mitochondria which reduces FAD to $FADH_2$. In mitochondria it is the oxidation of succinate that allows the reduction of FAD, whereas in the melanin granule it is likely the radiolysed water molecule that reduces FAD.

Since H_2 is the smallest non-polar molecule, it can easily diffuse throughout the body as a form of energy. With a solubility of about 0.0008 mol/L [6], the H_2 will remain in the body's fluid and diffuse throughout it without exiting the body. Granted, organisms with a larger surface area would be capable of generating more energy from sunlight. This makes sense of the conundrum of Kleiber's Law[7].

Kleiber's law, $E = M^{3/4}$, states that an animal's metabolic rate is approximately ¾ power of the animal's mass. For example, an elephant is about 10,000x more massive than a rat, but only requires 7,500x more energy than a rat, according to Kleiber's law. This law has plagued scientists for a while, because it is counterintuitive that larger organisms consistently have a more efficient metabolism. Phototrophic energy derived from the skin of animals may be the answer to this conundrum. Larger animals have larger surface areas, which may be the supplemental energy supply that causes a proportionally lower energy demand from food. The amount of energy derived from this photometabolic process can be calculated theoretically.

The sun radiates about 1.3 kW/m^2 at the earth's surface. This means that 1.3 kJ of energy per second is radiating on any given square meter of surface area. After taking into consideration the atmospheric dissipation, this radiation from the sun is actually around 1 kW/m^2 at habitable elevations. The average adult male has about $2m^2$ skin surface area. About half of the body can be radiated upon by sunlight at any given time, so this allows $1m^2$ of skin to absorb sunlight at a given time. The average daily energy demand is about 8,700 kJ per day. At 100% absorption and energy transfer efficiency, approximately 8,700 seconds, or 145 minutes, of 1 kW/m^2 radiation could theoretically suffice the daily energy demand of the human.

Note: these numbers are only applicable to a naked, very dark-skinned person. An untanned, pale, clothed person would not apply to these numbers. If you are pale, you should ideally get a considerable sun-tan before expecting sufficient amounts of energy from the sun.

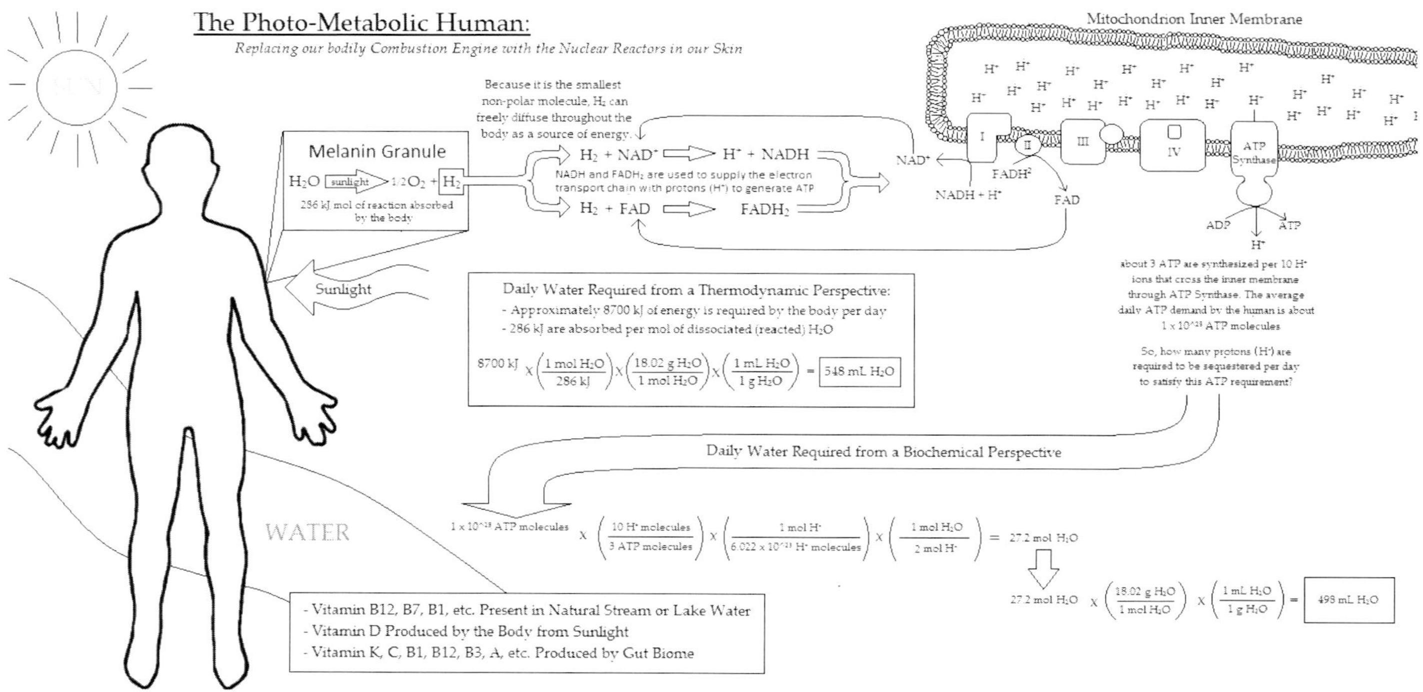
The Photo-Metabolic Human:
Replacing our bodily Combustion Engine with the Nuclear Reactors in our Skin
Melanin Granule
H_2O sunlight $1/2 O_2 + H_2$
286 kJ mol of reaction absorbed by the body
Sunlight
Because it is the smallest non-polar molecule, H_2 can freely diffuse throughout the body as a source of energy.
$H_2 + NAD^+ \Rightarrow H^+ + NADH$
NADH and $FADH_2$ are used to supply the electron transport chain with protons (H^+) to generate ATP
$H_2 + FAD \Rightarrow FADH_2$
Mitochondrion Inner Membrane
NAD^+
$NADH + H^+$
$FADH^2$
FAD
I
II
III
IV
ATP Synthase
ADP
ATP
H^+
about 3 ATP are synthesized per 10 H^+ ions that cross the inner membrane through ATP Synthase. The average daily ATP demand by the human is about 1 x 10^23 ATP molecules
So, how many protons (H^+) are required to be sequestered per day to satisfy this ATP requirement?
Daily Water Required from a Thermodynamic Perspective:
- Approximately 8700 kJ of energy is required by the body per day
- 286 kJ are absorbed per mol of dissociated (reacted) H_2O
$8700\ kJ \times \left(\frac{1\ mol\ H_2O}{286\ kJ}\right) \times \left(\frac{18.02\ g\ H_2O}{1\ mol\ H_2O}\right) \times \left(\frac{1\ mL\ H_2O}{1\ g\ H_2O}\right) = 548\ mL\ H_2O$
Daily Water Required from a Biochemical Perspective
$1 \times 10^{23}\ ATP\ molecules \times \left(\frac{10\ H^+\ molecules}{3\ ATP\ molecules}\right) \times \left(\frac{1\ mol\ H^+}{6.022 \times 10^{23}\ H^+\ molecules}\right) \times \left(\frac{1\ mol\ H_2O}{2\ mol\ H^+}\right) = 27.2\ mol\ H_2O$
$27.2\ mol\ H_2O \times \left(\frac{18.02\ g\ H_2O}{1\ mol\ H_2O}\right) \times \left(\frac{1\ mL\ H_2O}{1\ g\ H_2O}\right) = 498\ mL\ H_2O$
WATER
- Vitamin B12, B7, B1, etc. Present in Natural Stream or Lake Water
- Vitamin D Produced by the Body from Sunlight
- Vitamin K, C, B1, B12, B3, A, etc. Produced by Gut Biome

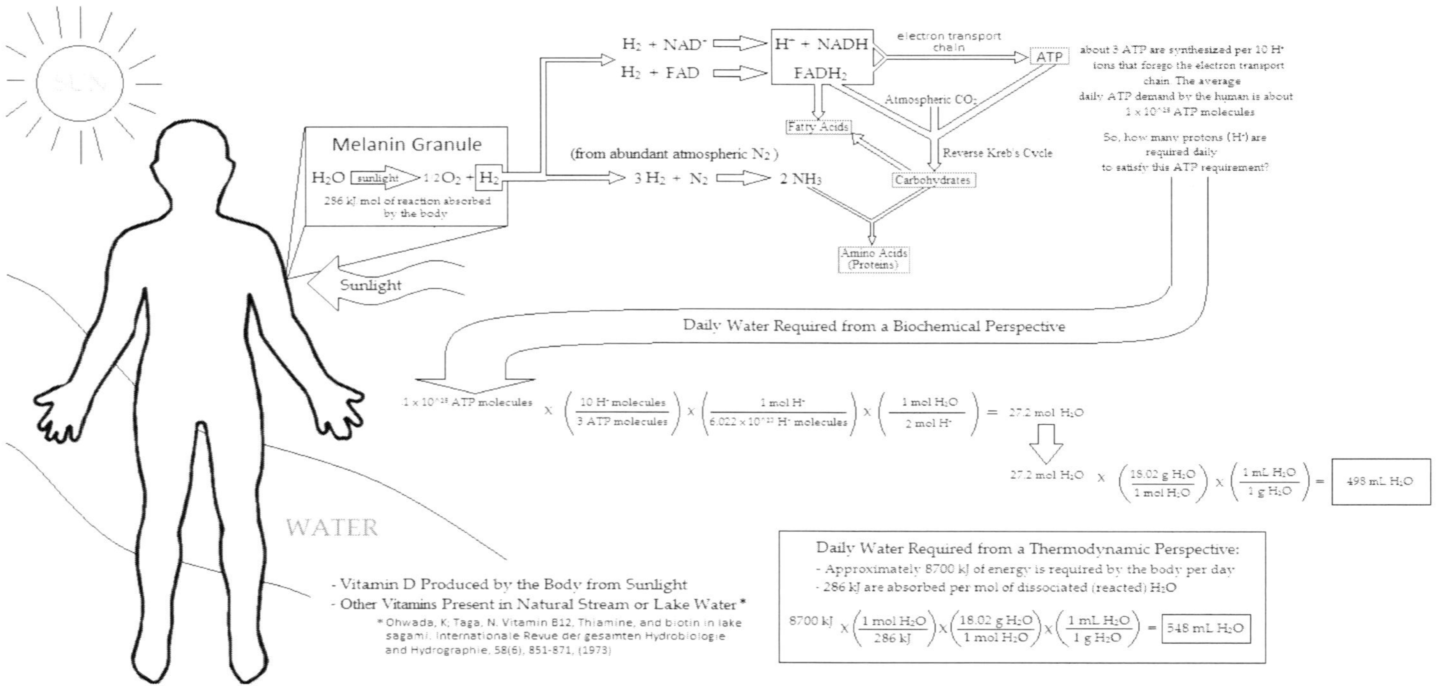
SUN
Melanin Granule
H2O sunlight 1/2O2 + H2
286 kJ mol of reaction absorbed by the body
Sunlight
H2 + NAD+
H2 + FAD
H+ + NADH
FADH2
electron transport chain
ATP
(from abundant atmospheric N2)
3 H2 + N2
2 NH3
Fatty Acids
Atmospheric CO2
Reverse Kreb's Cycle
Carbohydrates
Amino Acids (Proteins)
about 3 ATP are synthesized per 10 H+ ions that forego the electron transport chain. The average daily ATP demand by the human is about 1 x 10^25 ATP molecules
So, how many protons (H+) are required daily to satisfy this ATP requirement?
Daily Water Required from a Biochemical Perspective
1 x 10^25 ATP molecules x (10 H+ molecules / 3 ATP molecules) x (1 mol H+ / 6.022 x 10^23 H+ molecules) x (1 mol H2O / 2 mol H+) = 27.2 mol H2O
27.2 mol H2O x (18.02 g H2O / 1 mol H2O) x (1 mL H2O / 1 g H2O) = 498 mL H2O
WATER
- Vitamin D Produced by the Body from Sunlight
- Other Vitamins Present in Natural Stream or Lake Water*
*Ohwada, K; Taga, N. Vitamin B12, Thiamine, and biotin in lake sagami. Internationale Revue der gesamten Hydrobiologie and Hydrographie, 58(6), 851-871, (1973)
Daily Water Required from a Thermodynamic Perspective:
- Approximately 8700 kJ of energy is required by the body per day
- 286 kJ are absorbed per mol of dissociated (reacted) H2O
8700 kJ x (1 mol H2O / 286 kJ) x (18.02 g H2O / 1 mol H2O) x (1 mL H2O / 1 g H2O) = 548 mL H2O

100% efficiency is impossible, but Meredith and Riesz demonstrated that melanin absorbs energy at about 99.9% efficiency, and Herrera et al demonstrated that this absorbed energy is effectively used to split the water molecule. 286 kJ of energy is absorbed per mol of reacted water. This indicates that about 30.4 mols, or 548 mL, of reacted water catalyzed by melanin could theoretically satisfy the daily energy demand of the human. The skin is essentially acting as a nuclear reactor that feeds hydrogen and oxygen to a hydrogen fuel cell, the mitochondrion.

These numbers are also feasible from a biochemical perspective as well. ATP synthase generates about 3 ATP molecules per 10 H^+ that pass the inner membrane. 2 mols of H^+ are generated per mol of reacted water. The human requires around 10^{25} ATP molecules per day. This means about 27.2 mol, or 498 mL, of reacted water could suffice the daily ATP demand. It is important to note here that the enthalpic and biochemical demand are one and the same. These energy demands would not add upon each other but instead are different perspectives on the same estimate for daily energy demand. Theoretically, humans could live on nothing but nutrient-rich water, air and sunlight. Such a diet may be an effective therapy for cancer.

The Warburg Effect describes how most cancerous cells upregulate glucose consumption via glycolysis without a proportional increase in oxidative phosphorylation. If these cancerous cells need excessive glucose to maintain their malignancy, it would be intuitive that reducing available glucose could shunt the tumor. Research has already demonstrated that a ketogenic diet, by restricting carbohydrate (glucose) intake, is effective against tumors [8]. If this therapy is treated with the new metabolism discussed in this paper, it could potentially make the treatment much more effective. Since melanin, sunlight and water create energy that does not undergo glycolysis, but only oxidative phosphorylation, the Warburg Effect in cancerous cells could be starved by a water and sunlight diet.

Photons from the sun are not the only source of light for the human. Biophotons, ultra-weak photon emissions, range from 200-800 nm [9]. The energy of a photon is equal to Planck's constant times the speed of light divided by wavelength ($E_{photon} = hc/\lambda$). We know that water requires 286 kJ/mol to dissociate into H_2 and O_2. There are 286 kJ/mol of photons at a wavelength of 418 nm. Since melanin absorbs photonic energy at 99.9% efficiency, this means that one photon slightly below the wavelength of 418 nm will be able to split one water molecule. Biophotons have shown to be released at a rate of 1-1000 photons per second per cm^2 (photons/(s x cm^2))[10]. This translates to about 10-10,000 photons per cm^3 (centimeters cubed). The human body is about 70,000 cubic centimeters. This means that from any given moment the body's total emission of biophotons is about 700,000 to 700,000,000 per second for the average 70 kg person. This means hundreds of thousands of water molecules could be splitting per second to generate energy. This is a very substantial amount. Although, this number may be incorrect because the study that determined average biophotonic output was studying a 2-dimensional sample of tissue (cm^2) whereas the human body is a 3-dimensional being (cm^3). This internal energy emission is likely absorbed by neuromelanin in the substantia nigra and locus coeruleus, among other tissues which contain melanin. This means that the biophotonic energy emitted by the human could theoretically be converted into a significant amount of useful metabolic energy.

Energy from water fission could be stored as biomolecules as well. It is well known that the Kreb's cycle is reversible. The **reverse** Kreb's cycle and gluconeogenesis require NADH, $FADH_2$, and ATP which are ultimately generated from splitting the water molecule, and CO_2 which is readily available in the air. Ketone body synthesis from CO_2 and H_2 could be the physiologically preferable pathway for harnessing the energy as biomolecules, which would support the aforementioned therapy for the Warburg Effect. Even further, the superfluous amount of atmospheric nitrogen (N_2) that is regularly inhaled could react with the produced H_2 to form NH_3 which is required for protein synthesis. This is human photosynthesis. Furthermore, many nutrients and minerals are present in natural stream and lake water such as vitamin B12, biotin, thiamine, etc [11]. Our skin also produces Vitamin D naturally with sunlight.

In terms of having proper "nutrition", which may be completely altered with a new sort of metabolism like this, there are ways the body adapts to malnutrition. A group of South Africans called the Bantu, were found to have no access to dietary vitamin C. So, Andersson et al in "An investigation of the rarity of infantile scurvy among the South African Bantu", found that only a few people developed scurvy over the course of 5 years. How is this possible? Turns out that we have a gene that produces vitamin C, but it is inactive in most adults, likely due to the abundance of dietary vitamin C we consume. The human can make vitamin C when the nutrient becomes scarce.

Further proof of the adaptability of animals to nutritional scarcity comes from French Biologisit Louis Kervran. He found that Chickens that were completely deprived of calcium, a vital nutrient, but were given Potassium, were somehow generating eggs that contained calcium! Kervran then deprived the chickens of both potassium and calcium, and the chicken became calcium deficient. Kervran concluded that the chicken was somehow transmuting the abundant potassium into calcium [12]; which makes sense because potassium is adjacent to calcium on the periodic table. The scientific community is still baffled by this, and studying the mechanism of this is no easy task. Jean Paul Biberian in "Biological Transmutations" discusses this growing field and how biological organisms can transmute elements, which means to change the nuclear composition of elements. Hydrogen is a common component of transmutation [13], and thus hydrogen gas production via sun energy may be an important part of biological transmutation.

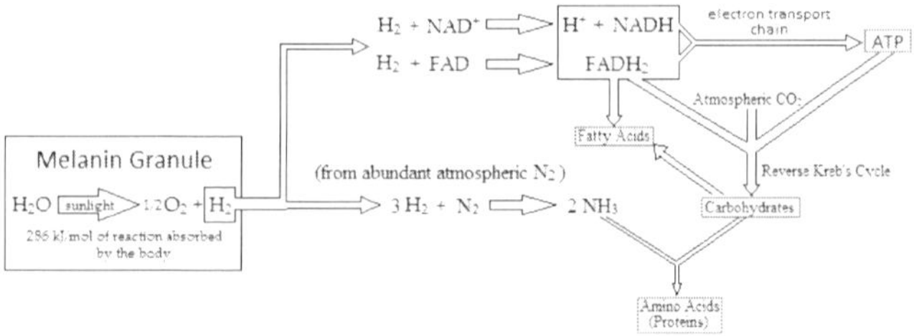

Figure 3. Biosynthesis supplied solely by H_2. H_2 can reduce NAD^+ and FAD to produce ATP via the electron transport chain. This ATP, with NADH, $FADH_2$ and atmospheric CO_2 is able to synthesis carbohydrates for storage via gluconeogenesis and the reverse Kreb's cycle. These Carbohydrates, along with additional NADH and $FADH_2$, can biosynthesis fatty acids for storage. NH_3 can be synthesis from atmospheric nitrogen and the synthesized H_2. This NH_3, along with various carbohydrates, can be used for amino acid synthesis

H_2 is also the best known anti-oxidant. This means that this photometabolic process is also very cytoprotective. Melanin present in the human retina produces H_2 and O_2 from sunlight; in fact, unusually high levels of O_2 in the retina are what gave Herrera et al the first inclination that melanin may be splitting the water molecule. H_2 therapy in the retina has demonstrated consistent therapeutic benefits; this H_2 exposure can be achieved naturally by sunlight activating retinal melanin. Considering all that has been said, maybe the sun isn't a carcinogenic monster after all.

You may be wondering about all the report that indicates sunlight exposure as a carcinogen. This is inconclusive [14]. Research Findings are now starting to blame sunscreen use for rising cancer rates [15]. In 1935 the risk of melanoma was 1 in 500. Now melanoma risk is 1 in 55. It is important to note that sunscreen was introduced and mass marketed around 1950. Since sunscreen only blocks ultraviolet radiation, other forms of radiation are allowed to be exposed to the human skin for abnormal periods. Think of getting sunlight like eating. A sunburn is our skin's natural way of saying "I'm full". Sunscreen eliminates this bodily indicator and allows radiation that is higher in energy than ultraviolet waves to bombard human skin for prolonged periods. Maybe we should let our natural sunscreen, melanin, do its job. As important as melanin is as a sunscreen, its biochemical implications via radiolysing the water molecule are much more important.

This new understanding of our skin will also increase the ability of athletes to train. The melanin in the skin allows a significant increase in the body's ability to do work, due to the increase in ATP availability. ATP generated in this way, unlike normal digestion, does not decrease the sympathetic nervous system which would allow more efficient training.

This may be the biochemistry that allowed an aboriginal tribe of Mexicans, the Tarahumara, to be able to run literally hundreds of miles. Their location near the tropic of cancer ensures they get adequate sunlight, so this may be their secret, which even they are unaware of. According to the book "Born to Run – A Hidden Tribe, Superathletes, and the Greatest Race the World has Never Seen", they were able to go on a 200 mile non-stop run over the span of a couple days. The journalist believed it was the type of shoes they were wearing, but human photosynthesis seems to be a more logical fit regarding how they were able to achieve such a feat.

Early humans were more in-tuned with this long-forgotten biochemistry. They were likely dark skinned (saturated with melanin) in order to absorb as much light energy as possible and drank water from natural sources. Abstinence from food, or fasting, has been praised for its beneficial effects since the dawn of writing and likely before written history as well. In our contemporary era, scientific report is beginning to understand the benefits of fasting. For example, rats that were forced to fast lived on average about 50% longer than rats that were given food ad libitum. This is amazing. This simple study effectively demonstrated the immense impact of abstaining from food. Also, when food is not being digested, blood-flow can be allocated away from the gut to increase the efficiency of other organ systems.

There is a group of people who call themselves breatharians who claims to live on nothing but air and sunlight. Until now, their ideas had no scientific backing so they were dismissed as insane by the mainstream. There is also a man in Ukraine, Nikolai Dolgoruky, who claims to have been living on water and sunlight for years. Researchers examined him and concluded that he was a healthy human.

Others have claimed to live solely from sunlight and water, including Hira Ratan Manek, Teresa Neumann, Giri Bala, and many more. Now we know how Nikolai and the others are able to survive; human photometabolism.

Nikolai also promotes the idea of “sun-gazing”, which involves looking at the sun during the low-intensity hours of sunrise and sunset. This also makes sense because there is a significant amount of melanin inside the eye. This could energize the not just the eye, but also the brain by transport of H_2 through the optic nerves. This can be VERY dangerous if practiced ignorantly. This book will not go over how to safely sun-gaze because it varies from person to person, although it is commonly directed to only do this during the first hour of sunrise or sunset.

You will know if you can perform a water fast or not. It is best to slowly transform your diet into a raw-only diet before you attempt a water fast. When I was younger I was not able to make it through the morning without eating. I would get overwhelmed with hunger. It was not until recently that I realized that mindset is very important. Mind over matter is no joke. Although, it is important to note that developing humans should be eating food.

Spiritual Implications

When revolutionary new information arises from scientific analysis, it can sometimes be matched to ancient holy texts that describe similar phenomenon. This is when truth is at its best. For example, prior to the big bang theory, it would surpass our intellect as to how God could make light of nothing, as proposed in Genesis 1:3. The Hubble telescope, among other experiments, demonstrated that all light and matter inexplicably originated from a single point and expanded outwards. How did biblical writers know about the big bang?

In the 19th century a man by the name of Andrew Crosse demonstrated that abiogenesis was possible. This means the creation of a biological organism by unbiological means. Crosse set up a vat containing silicates, water and an electrical current in an attempt to create artificial crystal formations. To his surprise, after a few weeks, little insects began growing on the surface of the minerals. Close examination determined that these bugs belonged to the genus acarus, and had a complete insect anatomy. These bugs grew from the silicate material, and were able to move and respond to external stimulus like the sun. People assumed that insect eggs infested his experimental setup, but the experiment was recreated by the London Electrical Society at the time. WH Weekes, of the Electrical Society, recreated Crosse's experiment and achieved the same results, with the same bugs. To prove that these bugs were not being created by biological means, Weekes roasted all his equipment above 400 degrees, much higher than any insect egg could withstand. He also doused the entire project in vaporous mercury, to ensure a biologically inviable environment. Crosse's experiment, and the successful recreation, are available in the book "Abiogenesis and Life from Dirt". This is amazing. Literally life from dirt. Even more interesting is that silicates, water, and electricity, the ingredients required to spawn these bugs, were the initial conditions in Genesis:

"In the beginning God created the heavens and the **earth**... and the Spirit of God was moving over the surface of the **water**. God said, "Let there be light," and there was **light**." (Genesis 1:1-3)

Earth: The earth's crust is comprised of about 70% silicates. This is the 1st ingredient in Crosse's experiment

Water: the 2nd ingredient in Crosse's experiment.

Light: Light is electromagnetic radiation. Electromagnetic radiation implies a source of "electricity". The 3rd ingredient in Crosse's experiment.

If someone can accidentally create life from these conditions, some more knowledgeable being could have possibly done much greater, complex acts of creation. Maybe we should suspend our disbelief momentarily and consider that this text is divinely inspired. The discovery of our skin's ability to act as a hydrogen fuel cell that requires only water and sunlight may be on the same magnitude as the big bang theory. The Buddha reached enlightenment after 49 days of fasting. Jesus Christ was offered limitless power and reign over the earth after 40 days of fasting.

Many other cultures have their own myths about deification through fasting, and the Bible is no exception:

"Meanwhile his disciples urged him, "Rabbi, eat something." But he said to them, "I have food to eat that you know nothing about."

(John 4:31-32)

What is this food that humans know nothing about? Could it be light, spiritual sustenance?

Jesus replied, "Very truly I tell you, no one can see the kingdom of God unless they are born again."

"How can someone be born when they are old?" Nicodemus asked. "Surely they cannot enter a second time into their mother's womb to be born!"

Jesus answered, "Very truly I tell you, no one can enter the kingdom of God unless they are born of **water** and the Spirit. Flesh gives birth to flesh, but the Spirit gives birth to spirit"

(John 3:4-6)

This spiritual birth that Jesus is eluding to is the soul's ascent. The fact that he mentions water and Spirit (the corporeal sun?) is very interesting in light of what has been discussed so far. **If ever a human soul were to escape the confines of materiality, a de-materializing lifestyle, such as the abstinence of food, would be an intuitive place to start.**

"Through the abandonment of desire, the deathless state is realized."

(Buddhism: Samyatta Nikaya xlvii 37)

It is said explicitly in the bible as to how the apostles were supposed to do the things Jesus did:

"...For truly I say to you, if you have faith as a grain of mustard seed, you will say to this mountain, 'Move from here to there,' and it will move. And nothing will be impossible for you. But this kind does not go out except by prayer and **fasting**."

(Matthew 17:20-21)

Is this revolutionary new science of water metabolism the legendary fountain of youth, perhaps? Ironically the conquistadors ravaged entire tribes looking for this secret which was intrinsic to their being the entire time. This theme also occurs in Greek philosophy:

"And if, my dear Socrates, Diotima went on, man's life is ever worth the living, it is when he has attained this vision of the very soul of beauty. And once you have seen it, you will never be seduced again by the charm of gold, of dress, or of sexual attraction; you will care nothing for the beauties that used to take your breath away and kindle such a longing in you, and many others like you, Socrates, to be always at the side of the beloved and feasting your eyes upon it, so that you would be content, if it were possible, to **deny yourself the grosser necessities of meat and drink**, so long as you were with it... And when you have brought forth and reared this perfect virtue, you shall be called the friend of god, and **if ever it is given to man to put on immortality, it shall be given to him**."

(Plato's Symposium 211-212)

in Genesis it was the act of eating that caused the fall of humankind into a world of death and decay:

" ...but of the tree of the knowledge of good and evil you shall not eat, for in the day that you eat from it you will surely die." (Genesis 2:17)

The tree of the knowledge of good and evil is likely symbolic for the material world in general, or the ego, but we can't refuse the fact that it was the act of eating that caused the fallen state of humankind. Following this act, it is said that the woman would now have to go through painful child-bearing as we know it today. Why did child-bearing only occur after the act of eating? It is because once we began to "eat" in the way we know it today, it would cause us to decay. Since death came after eating, procreation became required, otherwise the human race would become extinct after they died. Biological procreation was required because death came from eating.

Contemporary science supports this notion. The leading theory regarding why humans age is exactly that, our own metabolism. It is theorized that we die because of the metabolism of food; how ironic!

"The Mitochondrial Free Radical Theory of Aging (MFRTA) proposes that mitochondrial free radicals, produced as by-products during normal metabolism, cause oxidative damage. According to MFRTA, the accumulation of this oxidative damage is the main driving force in the aging process."

("The Mitochondrial Free Radical Theory of Aging": Sanz A, Stefanatos RK.)

Human photosynthesis is the keystone that the human intellect has been missing since the fall of Adam and Eve. The fountain of youth requires drinking water with an abstinence from food; such an abstinence likely stops us from aging.

The original human (fully expressed embodiment of God) was allowed to eat from the Tree of Life, but this may be eluding to a spiritual eating. Jesus separates "eating" into spiritual and material forms. In fact, the 5 loaves that fed the five thousand was not actual food, but some sort of spiritual sustenance:

Jesus asked, "You of little faith, why are you talking among yourselves about having no bread? Do you still not understand? Don't you remember the five loaves for the five thousand, and how many basketfuls you gathered? Or the seven loaves for the four thousand, and how many basketfuls you gathered? **How is it you don't understand that I was not talking to you about bread?**

(Matthew 16:8-11)

This "food" he is referring to is intrinsic to our being. Jesus further extrapolates on this matter in John 6:

Jesus answered, "Very truly I tell you, you are looking for me, not because you saw the signs I performed but because you ate the loaves and had your fill. **Do not work for food that spoils, but for food that endures to eternal life**, which the Son of Man will give you. For on him God the Father has placed his seal of approval."

(John 6:26-27)

When Jesus' fasts in the wilderness, it is preceded by his Baptism. What is baptism? You are baptized *with* water. After Jesus is baptized with water and the Spirit, he goes off into the wilderness on a 40 day fast, which by definition consists of only drinking water. Jesus is tempted with rulership over the world by the devil because he has gained such spiritual prowess by his 40 day fast. Jesus refuses the rulership, demonstrating that the heavenly realm transcends any material splendor. Think of it like an eternal dream in which you are in full control; you don't have to eat in dreams.

Jesus taught people how to activate their human potential, and this true potential is a spiritual birth without dependency on the material. Regardless of spiritual implications, human photometabolism is a new form of metabolism that can generate a significant amount of energy for the human from water and sunlight.

"This is the message we have heard from him and declare to you: God is light; in him there is no darkness at all. If we claim to have fellowship with him and yet walk in the darkness, we lie and do not live out the truth. But if we walk in the light, as he is in the light, we have fellowship with one another, and the blood of Jesus, his Son, purifies us..."

(1 John 1:5-7)

The savior said, "O blessed Thomas, of course this visible light shines on your behalf - not in order that you remain here, but rather that you might come forth - and whenever all the elect abandon materiality, then this light will withdraw up to its essence, and its essence will welcome it..."

(The book of Thomas the Contender by John D. Turner)

“Night shall be no more. They need no lamp nor the light of the sun, for the Lord God will give them light. And they shall reign forever and ever.”

(Revelations 22:5)

Bibliography

1) Valmalette et al. (2012). Light-induced electron transfer and ATP synthesis in a carotene synthesizing insect. Scientific Reports 2, 579
2) Dadachova et al. (2007). Ionizing radiation changes the electronic properties of melanin and enhances the growth of melanized fungi. PLoS ONE 2 (5), e457
3) Meredith, P; Riesz, J. (2004). Radiative Relaxation Quantum Yields for Synthetic Eumelanin. *Photochemistry and Photobiology*. 211-216.
4) Solis-Herrera et al. (2010). The unexpected capacity of melanin to dissociate the water molecule fills the gap between the life before and after ATP. *Biomedical Research* 21(2), p 224-226.
5) Rosania, Gus. (2005). Mitochondria Give Cells a Tan. *Chemistry & Biology* 12(4), 412-413.
6) GWC Kaye and TH Laby, "Tables of Physical and Chemical Constants" 15th ed., Longman, NY, 1986, p. 219.),
7) Aliev et al. (2013). Human photosynthesis, the ultimate answer to the long term mystery of

Kleiber's law or E = $M^{3/4}$: Implication in the context of gerontology and neurodegenerative diseases. Open Journal of Psychiatry 3, 408-421.

8) Seyfried et al. (2009). Targeting energy metabolism in brain cancer through calorie restriction and the ketogenic diet. 5:1, S7-15
9) Dalmau-Santamaria, I. (2013). Biophotons: A modern interpretation of the traditional "qi" concept. Revista Internacional de Acupunctura. 7:2, 56-64
10) Ives JA, van Wijk EPA, Bat N, Crawford C, Walter A, et al. (2014) Ultraweak Photon Emission as a Non-Invasive Health Assessment: A Systematic Review. PLoS ONE 9(2): e87401
11) Ohwada, K; Taga, N. Vitamin B12, Thiamine, and biotin in lake sagami. Internationale Revue der gesamten Hydrobiologie and Hydrographie, 58(6), 851-871, (1973)
12) Biological Transmutations. C.L. Kervran.
13) Xu et al. (2009). Effects of hydrogen and helium produced by transmutation reactions on void formation in copper isotopic alloys irradiated with neutrons. Journal of Nuclear Materials. Vol 386-88, p363-366.

14) Levell et al. (2009) Melanoma epidemic: a midsummer night's dream? Br J Dermatol. 161(3), 630-634.
15) Westerdahl et al. (2000). Sunscreen use and malignant melanoma. International Journal of Cancer. 87(1), 145-150.

Made in United States
North Haven, CT
04 November 2026

21769079R00021